During the night, a yellow-necked field mouse (cover) leaves its nest to look for food.

A saw-whet owl looks out of its home in a tree.

by Judith E. Rinard

BOOKS FOR YOUNG EXPLORERS
NATIONAL GEOGRAPHIC SOCIETY

 LIBRARY OF CONGRESS CIP DATA: PAGE 32

In the evening, as it grows dark,
raccoons go down to a river.
They are hunting for food in the water.
Raccoons sleep during the day.
But at night they wake up and are hungry.
A baby raccoon slips, and almost falls.

Another raccoon catches
a frog and eats it.
There are many animals,
like raccoons, that come out
when the sun goes down.
They are creatures of the night.

During the day,
the leopard rests in a tree.
When evening comes,
it climbs down and
hunts for food.
The leopard can see
in the dark.
And it can move quietly
on its soft, padded paws.
Sometimes, other animals
do not hear
the leopard coming.
Then, the leopard
may catch them
and kill them quickly
with its long, sharp teeth.

Many animals call to each other in the darkness. As night ends, a coyote sits in the snow and calls to another coyote. Have you ever heard any of the sounds that animals make at night?

A mockingbird sings in a tree.
It sings in the evening
and far into the night.

BAHAMA MOCKINGBIRD

SNOWY TREE CRICKETS

COQUI FROG

A male cricket chirps
by rubbing his wings together.
When the female hears this song,
she may crawl down the twig to him.
The male frog puffs up his throat
and calls to his mate.

At night a tree sparkles with fireflies.
They look like lights on a Christmas tree.
Fireflies find their mates by flashing their tiny lights.
The male flashes the light at the end of his body.
The light blinks on and off many times.
If the female sees the signal, she usually blinks right back.
The light of a firefly makes a bright spot on a leaf.

The railroad worm shines its lights when something frightens it.
When the spots along its body shine,
the lights may scare hungry animals away.

RAILROAD WORM

A young kit fox hears a noise and stops.
It stays very still and listens quietly.
A serval cat also listens with its ears perked up.
It hunts at night and listens for
animals moving in the dark.
Some night hunters can hear nearly every sound.

A mouse listens for sounds of danger.
When a sound frightens it, the mouse dashes away.

KIT FOX

AUSTRALIAN HOPPING MOUSE

SERVAL

BUSHBABIES OR GALAGOS

SQUIRRELFISH

Two bushbabies peer into the darkness.
With their big, round eyes, these animals of Africa
look for insects in a tree.
Many animals that hunt at night have large eyes
that help them see well at night.

The squirrelfish hunts in the ocean after the sun sets.
It searches for tiny sea animals in the dark water.
A small owl looks out of its nest in a tree.
The furry loris has eyes that seem to pop out of its head.
In the daytime it curls up in a tree.
At night, when it wakes up,
the loris feeds on insects, fruit, and bird eggs.

SCREECH OWL

SLOW LORIS

CLIMBING MOUSE

This mouse is looking for seeds to eat.
As soon as it gets dark,
the mouse creeps out of its nest in the grass.

A hungry owl has seen a white-footed mouse.
The owl swoops down to catch it.
The mouse does not hear the owl
because it flies without making any noise.
The owl catches the mouse in its sharp claws
and carries it to a tree.
Owls can find little mice on the blackest night.
These birds have keen ears that can hear
a mouse scurrying in the grass.
They also have sharp eyes.
Owls are very good hunters.

SAW-WHET OWL

SAW-WHET OWL

TAWNY OWL

There are many kinds of owls. Most of them hunt only at night, when the animals they eat are moving about.
A tawny owl flaps its wings as it lands.
The great horned owl sits very still all day long.
A barn owl flies out of its nest, where it rests during the day.

GREAT HORNED OWL

BARN OWL

BROWN KIWI

TAWNY FROGMOUTH

BLACK-CROWNED NIGHT HERON

Flying low, a night heron carries a fish home. To catch fish, the heron stands very still in the water until one swims by.
The kiwi has nostrils at the tip of its long bill.
It finds worms by smelling them in the ground.
This brown, shaggy bird cannot fly because its wings are very tiny.
The frogmouth is a bird that catches insects in its bill.
When insects come near, the bird zooms down and snaps them up.

HORSESHOE BAT

Bats are not birds. They are the only mammals that can fly.
This horseshoe bat searches for insects to eat.
It catches insects in the air
by scooping them up in its wide wings.
All day, four bats, called flying foxes, hang upside down in a tree.
In the evening, they will fly off and look for fruit to eat.
Long-nosed bats feed on nectar from flowers.
One laps up the sweet juice with its long tongue.
The other bat pokes its head deep inside a cactus flower.
Maybe it is after the very last drop of nectar.

FLYING FOXES

SANBORN'S LONG-NOSED BATS

A flying squirrel leans out of a hole in a tree.
It leaves its cozy nest when the sun goes down.
Then it leaps from a high branch and spreads the flaps of skin between its front and hind legs. The flaps are like a little parachute.
They help the squirrel glide down through the air.
Two squirrels peek out of their home.
A third squirrel sits outside and nibbles on a nut.

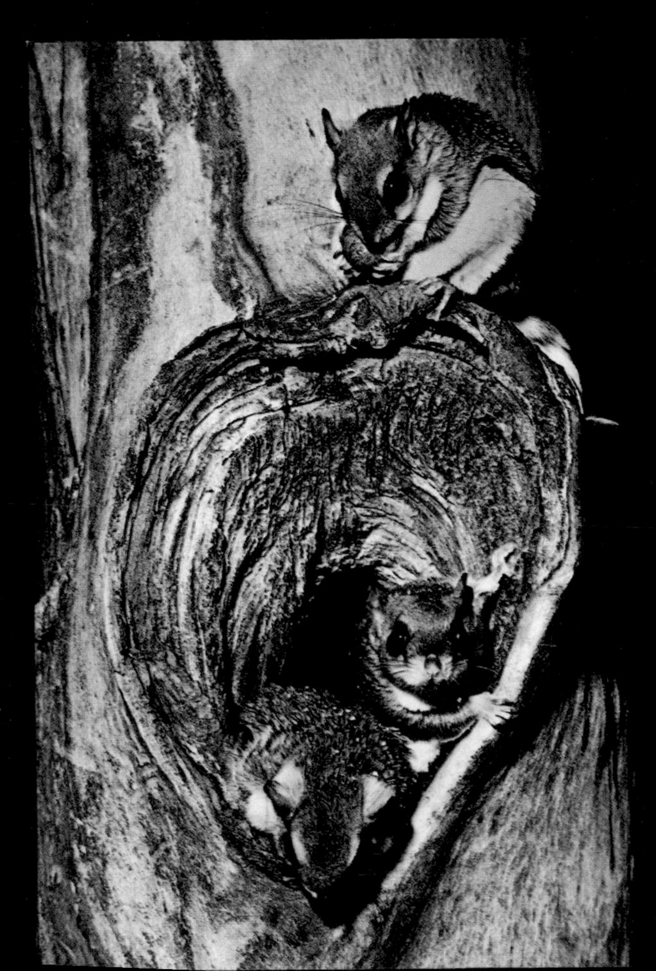

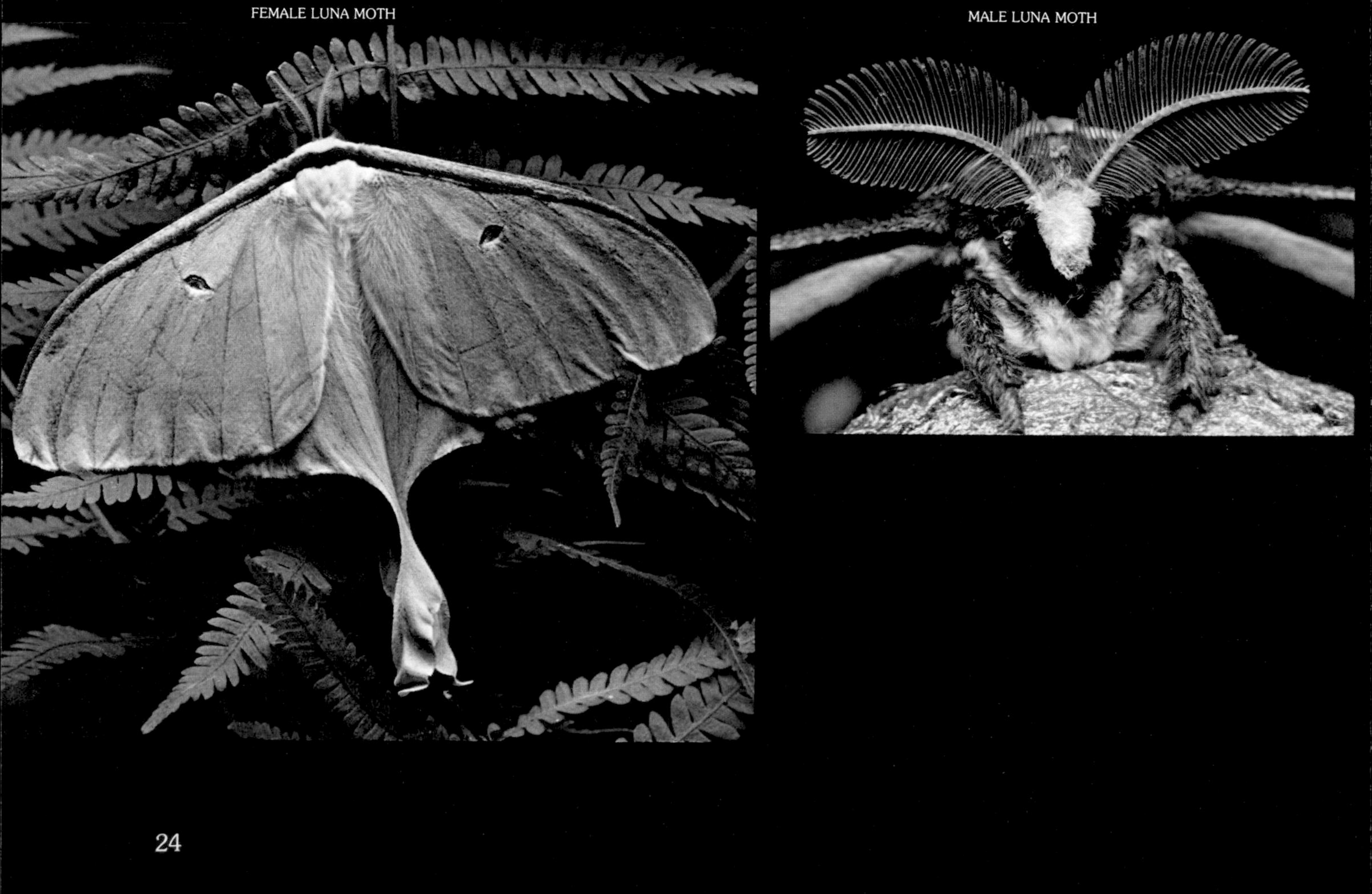
FEMALE LUNA MOTH

MALE LUNA MOTH

HAWKMOTH AT A RANGOON CREEPER

In the ocean, a moray eel swims out of its den. This large, powerful eel has two tubes on the end of its snout. The tubes help the eel smell and find fish in the dark waters.

TESSELLATED REEF EEL

Most shrimp feed at night.
But the red rock shrimp
also hunts during the day.

RED ROCK SHRIMP

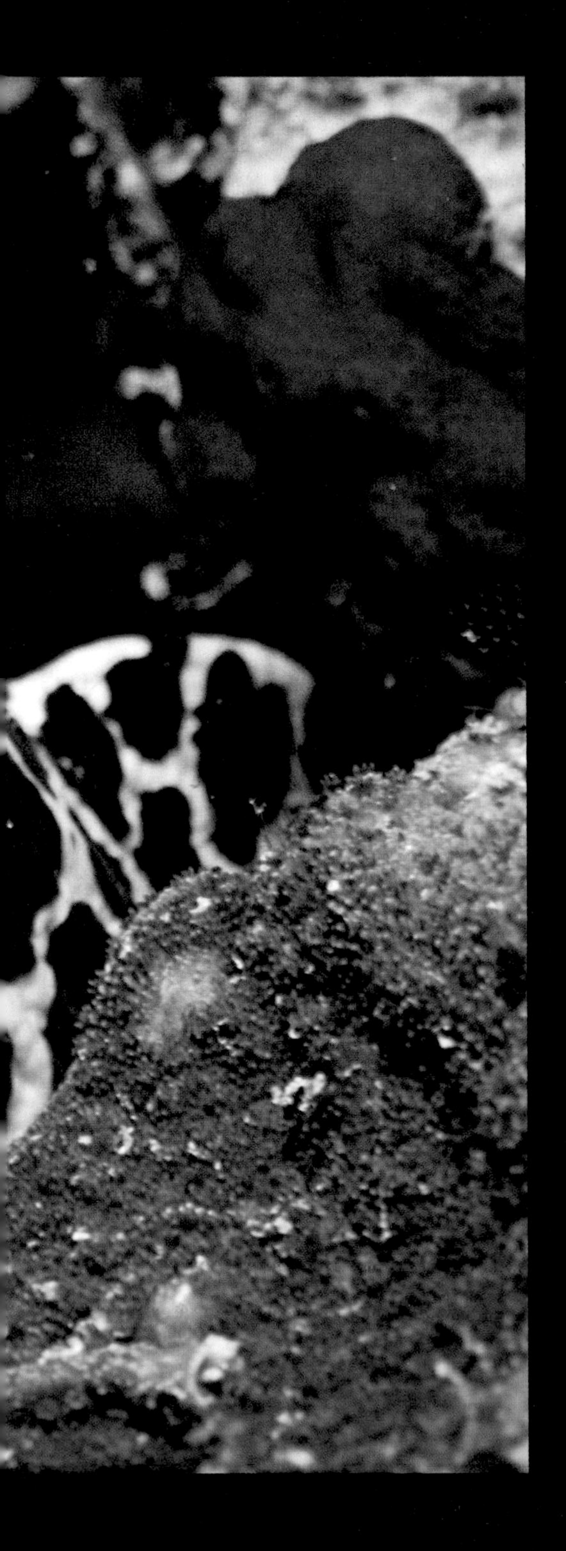

In the daytime, this basket star
looks like a dead plant.
But at night, it unfolds its arms
and catches tiny sea creatures.

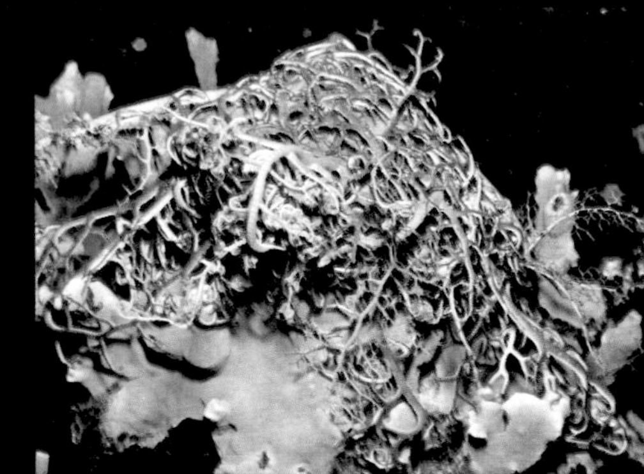
BASKET STAR
OR GORGON'S HEAD

GARDEN SNAIL

BROWN SLUG

Snails and slugs also look for food at night.
Many slugs in the ocean
have bright, beautiful colors.
The purple sea slug breathes
through its orange gills.
The garden snail and brown slug live on land.
They have soft, moist bodies.
So if they stayed out during the day,
the sun would dry them out.
The sea snail has two long feelers.
It uses them to find its way
along the dark bottom of the sea.

NUDIBRANCH (LEFT) AND SUNDIAL (BELOW)

A mother opossum carries her babies to a tree. The night is almost over. Day is coming, and it is time for the opossum family and other creatures of the night to go to sleep. All night, while you are sleeping, many animals are wide awake.

Isn't the world of the night
a busy place?

Published by The National Geographic Society
Robert E. Doyle, *President;* Melvin M. Payne, *Chairman of the Board;*
Gilbert M. Grosvenor, *Editor;* Melville Bell Grosvenor, *Editor Emeritus*

Prepared by
The Special Publications Division
Robert L. Breeden, *Editor*
Donald J. Crump, *Associate Editor*
Philip B. Silcott, *Senior Editor*
Cynthia Russ Ramsay, *Managing Editor*
Sallie M. Greenwood, *Researcher*
Wendy G. Rogers, *Communications Research Assistant*

Illustrations
Geraldine Linder, *Picture Editor*
Jody Bolt, *Art Director*
Suez B. Kehl, *Assistant Art Director*
Drayton Hawkins, *Design and Layout Assistant*

Production and Printing
Robert W. Messer, *Production Manager*
George V. White, *Assistant Production Manager*
Raja D. Murshed, June L. Graham, Christine A. Roberts, *Production Assistants*
John R. Metcalfe, *Engraving and Printing*
Debra A. Antonini, Jane H. Buxton, Suzanne J. Jacobson, Cleo Petroff,
Katheryn M. Slocum, Suzanne Venino, *Staff Assistants*

Consultants
Dr. Glenn O. Blough, Peter L. Munroe, *Educational Consultants*
Edith K. Chasnov, *Reading Consultant*

Illustrations Credits
Zig Leszczynski, Animals Animals (1, 15 right); Russ Kinne, Photo Researchers (2-3, 23 bottom); Thase Daniel (2 bottom, 17 top); Karl H. Maslowski, Photo Researchers (3 bottom); Oxford Scientific Films Ltd., Bruce Coleman Inc. (4-5 top); George H. Harrison, Grant Heilman (4 bottom); M. P. Kahl, Bruce Coleman Inc. (5 bottom); Ralph Hunt Williams (6-7); A. Sprunt, National Audubon Society Collection/PR (7 top); Lois Cox (7 middle); Fran Hall, Photo Researchers (7 bottom); Dr. Ivan Polunin, Natural History Photographic Agency (8 top left); Runk/Schoenberger, Grant Heilman (8 top right, 29 top right); Robert F. Sisson, National Geographic Natural Science Photographer (8 bottom); Dr. Ivan Polunin, Bruce Coleman Inc. (9); Tom Myers (10, 11 left, 29 top left); Michael Morcombe, Natural History Photographic Agency (11 right); Volkmar Wentzel (12-13); Bates Littlehales, National Geographic Staff (12 bottom, 26-27, 27 top right, 28-29); Fred J. Alsop III, Bruce Coleman Inc. (13 left); Bruce Coleman Inc. (13 right); P. Ward, Bruce Coleman Inc. (14-15 top); Ron Austing, Bruce Coleman Inc. (14-15 bottom); Stephen C. Dalton, Natural History Photographic Agency (16-17); Hans Reinhard, Bruce Coleman Inc. (17 bottom); Edmund Appel, Photo Researchers (18-19 top); M. F. Soper, Bruce Coleman Inc. (18 top); A. W. Ambler, National Audubon Society Collection/PR (18-19 bottom); S. C. Bisserot, Bruce Coleman Inc. (20-21); Bruce Dale, National Geographic Photographer (21 top); Nina Leen (21 bottom); Stouffer Productions, Animals Animals (22); Stouffer Productions, Bruce Coleman Inc. (23 top); James H. Carmichael, Jr. (24 left); Anthony Bannister, Natural History Photographic Agency (24 right, 25); Walter A. Starck II (27 center); Robert E. Schroeder (27 bottom); Paul A. Zahl, National Geographic Staff (29 bottom); Jack Dermid (30-31); Dale R. Thompson and George R. Dodge, Bruce Coleman Inc. (32).

Cover Photograph: Stephen C. Dalton, Natural History Photographic Agency
Endpapers: C. C. Lockwood

Library of Congress CIP Data
Rinard, Judith E   Creatures of the night.   (Books for young explorers)
SUMMARY: Describes the after-dark activities of many nocturnal animals.
1. Nocturnal animals—Juvenile literature.   [1. Nocturnal animals]   I. Title. II. Series.
QL755.5.R56 591.5 77-76968 ISBN 0-87044-214-4

As the sun comes up
a snail creeps away.
It will soon be day,
and the snail will find
a shady spot to rest.